26th Annual International Conference on Mechatronics and Machine Vision in Practice (M2VIP 2019)

Toowoomba, Australia
3-5 December 2019

ISBN: 978-1-7138-0209-9

Printed from e-media with permission by:

Curran Associates, Inc.
57 Morehouse Lane
Red Hook, NY 12571

Some format issues inherent in the e-media version may also appear in this print version.

Printed with permission by Curran Associates, Inc. (2020)

Mechatronics and Machine Vision in Practice
FOES, University of South Queensland
Toowoomba QLD 4350 Australia

johnbill@usq.edu.au

Additional copies of this publication are available from:

Curran Associates, Inc.
57 Morehouse Lane
Red Hook, NY 12571 USA
Phone: 845-758-0400
Fax: 845-758-2634
Email: curran@proceedings.com
Web: www.proceedings.com

Welcome

Once again we would like to welcome you to the annual conference on Mechatronics and Machine Vision. This is the 26th in the series and we hope that it will be as successful and enjoyable as the previous ones. This conference has this year returned to Toowoomba, where visitors can gaze from the heights of the Great Dividing Range across the Lockyer valley almost to Brisbane and probably meet the kookaburra featured as our logo.

As always, the papers presented at this conference have survived a rigorous refereeing process, first as abstracts to check that they really reflect 'in practice' and then again as full manuscripts. The survivors have been given a final polish in the light of the comments from the reviewing process and most will be revised yet again for inclusion in a second book in the series published by Springer Verlag.

Many papers fell by the wayside. The survivors should stimulate the delegates to constructive discussion. Several of our friends from previous conferences will be here and we extend a warm welcome to them.

With three evenings of social events, there should be plentiful opportunities to meet old friends, make new acquaintances and maybe plot out some international collaborative projects.

With all very best wishes,

John Billingsley and Peter Brett

Mechatronics and Machine Vision in Practice 2019

Toowoomba, Queensland Australia, December 3-5

Tuesday 3rd December

9:25 Opening Welcome on behalf of USQ

9.45 - 10.45 Session1 Vision and optical sensing

The design of optical sensing needles for tactile sensing and tissue identification at needle tip

Zonglai Mo1,4, Ce Zhang2,4, Weiliang Xu2,4, and Neil G. R. Broderick3,4
1. School of Mechanical Engineering, Nanjing University of Science and Technology, China
2.Department of Mechanical Engineering, The University of Auckland
3.Department of Physics, The University of Auckland, New Zealand
4.The Dodd-Walls Centre for Photonic and Quantum Technologies, New Zealand

Abstract. Force sensing in minimally invasive surgery is a challenge for surgeons in practice and in training, and for the application of invasive devices. Within magnetic resonance (MR) working environments the solutions are limited. Optical sensors are a promising choice as they are immune to MR influence. This paper proposes optical sensing solutions for needle-tip applications of needle-tip tactile sensing and bio-tissue identification, based on interference intensity modulated Fabry-Perot interference (FPI) principle. The design of the sensor and its optical circuit are detailed described. Then, simulations, phantom tests, ex-vivo and in-vivo animal tests have been conducted to verify the design. During the experiments, sensor properties and environmental influence factors on the sensor were investigated. The results show that the designed needles are capable of identifying tissue layers and types via analyzing force signals during the insertion process in real time mode in a temperature-changeable environment, with effective temperature compensation designs.

Keywords: Needle sensing, Tissue identification, Intensity modulation, Force sensor

Object Detection on Train Bogies Using Structured Light Scanning

Tangwen Yang 1*, Yantao Sun 1, Xiaoqing Cheng 2, Honghui Dong 2, Yong Qin 2
1Institute of Information Sciences, School of Computer and Information Technology
2the State Key Laboratory of Rail Traffic Control and Safety
Beijing Jiaotong University, Beijing 100044, China
Email: twyang@bjtu.edu.cn
Abstract. In this paper, a non-contact technology based on structured light scanning is introduced to inspect adhered substances on train bogie, and compute the volume of substances. The scanning system consists of a linear laser and a camera. The images recorded by the system is used to reconstruct the 3D model of a train bogie. The iterative closest point (ICP) algorithm is introduced to align the reconstructed model to an initial background model. Then a point cloud difference is obtained by subtracting the aligned point cloud from the initial model. Using the numerical integration, the vol-ume of adhered substances is calculated from the data of the point cloud difference. The experimental results on a train bogie setup show that the structured light scanning method can reconstruct the 3D model of the bogie and accurately measure the volume of irregular adhered substances.
Keywords: Object Detection; Volume Measurement; Train Bogies; Structured Light Scanning; 3D reconstruction

Method for detecting breaking rate of *ganoderma lucidum* spore powder based on machine vision

Shanling Ji, Zhisheng Zhang, Zhijie Xia and Ying Zhu

Abstract. Ganoderma lucidum spore powder is a traditional Chinese medicine that needs to be broken to function more effectively. The traditional method of calculating the rate of destruction depends on professional experience and is inefficient. In this paper, the detection platform of Ganoderma lucidum spore powder breaking rate was developed based on machine vision. Firstly, the foreground area of spore microscopic image was extracted by local gray threshold processing. The overlapping spores would be separated, and then the characteristics of foreground area were extracted and classified by machine learning. The location and number of unbroken spores were then masked in the original image. The breaking rate was finally calculated according to the number of unbroken Spores sampled before and after processing of the same batch of Ganoderma lucidum spore powder. Some experiment was conducted and the accuracy of platform was about 95%.
Keywords: Ganoderma lucidum spore powder · Breaking rate · Computer vision · Local gray range threshold processing

11:10-12:30 pm Session 2 Agriculture Applications 1

Identifying research priorities – keys to successful mechatronic and data-dependent applications in the Australian sugarcane industry

BL Schroeder1, TA Jensen1, JH Panitz2, AW Wood3
1University of Southern Queensland, Toowoomba, QLD, Australia;
2Sugar Research Australia, Bundaberg, QLD Australia;
3Tanglewood Agricultural Services, Millaa Millaa, QLD, Australia
Bernard.Schroeder@usq.edu.au

Abstract.
Sugarcane farmers are generally striving to adopt sustainable in-field practices aimed at profitable and environmentally responsible production. Several changes to the sugarcane cropping systems have occurred in Australia in the past few decades. However, novel mechatronic and/or data-dependent applications are now available for inclusion in on-farm management strategies that are aligned with Precision Agriculture (PA) in typical Australian sugarcane production systems. We highlighted where R&D opportunities exist within the different stages (covering the fallow, plant crop, ratoon crops and harvesting) of a typical semi-irrigated rainfed farm with a one-year production cycle. These were ranked according to a simple system from currently available/being adopted to requiring R&D effort. The R&D opportunities classified according to complexity and estimated timeframe for successful outcomes identified a relatively large number of potential projects. Some opportunities were recognised as being possible in the shorter term and could be achieved using existing knowledge and technologies. Others were identified as being highly complex and needing relatively long R&D timeframes to produce workable outputs. In terms of the latter, two opportunities related to the development of (i) automated and robust sugar sensors to determine commercial cane sugar (CCS) at strategic points in the crop at harvest, and (ii) automated/semi-automated in-field robotic soil test equipment. Irrespective of the level of complexity, we suggest that comprehensive evaluation is needed to ensure the adoption of appropriate, reliable and robust mechatronic and/or data-dependent application for use in sugarcane production. Only then will these applications/technologies be ready for acceptance as part of industry BMPs and routinely adopted by growers with confidence and enthusiasm.
Keywords: Sugarcane, in-field practices, novel mechatronic and data-depend-ent applications

ROS for Agricultural Robotics: A Common Middleware Perspective

Anand Kumar Pothula and Craig Baillie
Centre for Agricultural Engineering, University of Southern Queensland, QLD 4350, Australia

Abstract.
Concerns over food security have increased over the last decade as the world food demand is predicted to increase by 70% (2050). Researchers are working towards the alternatives for increased food production with limited or diminishing land and water resources. Agricultural robots can increase crop yields by efficiently utilising inputs and resources while lowering the production cost. Robots are very complex in nature and multidisciplinary skills are required to build a complete system. Middleware provides an abstraction layer between low level hardware abstraction and device control to high level software application. However, several middleware development tools in practice make it very difficult to share software and re-implement code. Robots are heterogeneous in nature and finding a common middleware approach is very difficult. ROS is becoming increasingly popular as a common middleware for agricultural robots to

address this issue.

Keywords: ROS, Middleware, Agricultural Robots, Common Sensor Platform, Modular, Real-time

Vision-based Detection and Counting of In-field Fruits with Fully Convolutional Network

Jinky G. Marcelo1,2, Joel P. Ilao1, Macario O. Cordel II1
1 De La Salle University, Manila, Philippines
2 Central Mindanao University, Maramag, Bukidnon, Philippines
jinkymarcelo@cmu.edu.ph, joel.ilao@dlsu.edu.ph, macario.cordel@dlsu.edu.ph

Abstract. In agriculture, camouflage, i.e. fruits are occluded due to high color similarity to its environment, is one of the problems of fruit detection and counting for yield estimation. In this paper, we developed a system that automatically detects and counts fruits from images in an unstructured environment with heavy occlusion and high color similarity. We built two datasets by capturing the natural settings of the fruit and fine-tuned various deep convolutional neural networks with region proposal network for a very challenging task of detecting heavily occluded fruits with high color similarity against the background. The evaluation results show an average precision of 0.922 for chili pepper dataset and 0.966 for bell pepper dataset. The result of this network can be integrated into an application for efficient management of agricultural farms, thereby contributing to precision agriculture.
Keywords: Fruit detection and counting, yield estimation, camouflaged object detection, deep convolutional neural networks.

A pragmatic automated site-specific weed management system using unmanned aerial vehicle

Bruen Smith, Livia Faria Defeo, Troy Arnold Jensen
Centre for Agricultural Engineering, University of Southern Queensland, Toowoomba, 4350

Abstract
There is currently a need to optimize herbicide application technology to minimize herbicide use and resistance. To more fully utilize precision weed spray equipment a system has been developed implementing the four-step closed loop precision agriculture scheme. The aim being, to create a pragmatic system that utilizes existing precision spray and consumer grade UAV technology. The use of consumer grade UAV technology limited the system being developed to rely RGB imagery. Therefore, an algorithm utilizing the Excess green vegetative index was developed. This algorithm has been developed to minimize the detail and number of polygons in the resulting spray maps while maintaining the efficiency of the system. Ground truthing of the system reviled a hit rate of 92.6% and a potential herbicide saving of 44.6%.
Keywords. Machine vision, Precision agriculture, Unmanned Aerial Vehicle, Weed management

1.45 - 2.30 Session 3 -Keynote 1

2.30 - 3.10 Session 4 Robotics and Devices 1

Fabrication and Characterization of 3D Printed Microfluidics

Swapna A. Jaywant, Muhammad Asif Ali Rehmani, Tanmay Nayak
and Khalid Mehmood
Massey University, Department of Mechanical and Electrical engineering,
SF&AT, Auckland-0632, New Zealand k.Arif@massey.ac.nz

Abstract. Microfluidics is a rapid, high throughput and cost-effective means of processing small quantities of fluids through micro-scale channels. Several techniques are available today for the manufacturing of microfluidic channels such as injection moulding, soft-lithography, and paper microfluidics. Due to the advancements in modern additive manufacturing methods, 3D-printing (3DP) has been shown as a promising platform for the fabrication of microfluidic devices. In this paper, we explore the possibility of using fused deposition modelling (FDM) and selective laser sintering (SLS) technology for internal features of the microfluidic devices. These technologies have been compared in terms of their ability to print internal features to fabricate microfluidic devices. The outcome of the study provides a comparison of the minimum possible diameter size, fluid flowrate, and leakage through the internal features of the microfluidic devices.
Keywords: Microfluidics, 3D Printing, Characterization

A Modified Bresenham Algorithm for Control System of FDM Three-dimensional Printer

Ke Yu, Zhisheng Zhang*, Zhiting Zhou, Min Dai

Abstract. The essence of the FDM (Fused Deposition Modeling) three-dimensional printing control technology is that the extrusion nozzle completes coordinate point-to-point motion by stepper motors according to a pre-specified trajectory from G-codes. Bresenham algorithm is often applied in the occasions where the a straight trajectory needs to be interpolated. A modified Bresenham algorithm is introduced in this paper. On the basis of the conventional Bresenham algorithm, the modified method tried to improve the efficiency by the simplification of the decision process for each step advanced and the combination of motions of multiple steps. In order to achieve the improvement, a transformation of the decision formula is made and then the modified formula could be applied to predict the position of a plurality of interpolation points by one decision at a time. Furthermore, taking into account the principle of linear motion generation in the control system of FDM 3D printer, the paper gives preconditions and specific procedures to use the new algorithm on a MCU based 3D printer control system.

Keywords: Modified Bresenham algorithm · FDM Three-dimensional printer ·MCU

3:35 - 4:35 pm Session 5 Robotics and Devices 2

Design and experimental study on the self-balancing foot device

Rui Peng, Liang Han*
School of Mechanical Engineering Southeast University, Nanjing 211189, China
Corresponding author E-mail: melhan@seu.edu.cn

Abstract. Stabilizing and aligning tables often presents a problem when in use. In this paper, a foot device that can finely adjust the horizontal height within a certain range is designed. The device can automatically adjust the height of the foot according to the ground flatness to correct wobbly tables, and an external force can be applied to finely change the height of the table, so that multiple pieces of table can be perfectly aligned. The single foot is in the form of a piston cylinder, and each foot is connected with others by an elastic rope assembly. The height of the foot is changed through the telescopic movement of the piston in the cylinder. There is a spring inside the piston cylinder to bear the main load, and the brake element is designed to ensure that the height of the foot cannot change under the load of a certain range, so that the height of the table remains stable during normal use. In addition, the mechanical analysis of the foot device structure is carried out to explore the relationship between the piston telescoping and the load it carries, to optimize its bearing capacity and improve stability. The prototype of the foot device is fabricated, and the test system is built, which includes displacement sensor, pressure sensor, pressure loading device and controller. Then related experimental studies are carried out on the prototype, including testing the adjustment range, the height change of the foot under no load in a long time and the height change under different loads. The experiment results show that the adjustment range and stability fulfills requirements.
Keywords: foot device; balanced height; brake element; mechanical modeling

Dynamic Characteristics Analysis and Optimization Design of Cross-beam Assembly in 3D Printer

Weijie Chu, Fang Jia*, Limiao Gu, Xiaolong Liu
Department of Mechanical Engineering Southeast University Nanjing, 211189, China
Corresponding author E-mail:chuweijie138@163.com

Abstract. As the carrier of sprinkler movement of FDM 3D printer, the movement accuracy of the beam directly determines the printing accuracy. In this paper, the beam assembly of a super large size FDM3D printer is analyzed. Firstly, the modal analysis is executed, and its 12-order modes and its dominant factors are obtained. The modal test is implemented to verify the validity of the model. Then, the harmonic response analysis of the cross-beam assembly is implemented, and it is found that the displacement values are higher than the printer precision requirements. Finally, using Workbench to achieve topology optimization and size optimization of the cross-beam component, a new cross-beam structure is obtained, and the first-order natural frequency is increased by 9.4% while the cross-beam component mass is reduced by 43.6%. The research results can provide a theoretical basis for the design and optimization of the cross-beam assembly of FDM 3D printer.
Keywords: FDM 3D printer; cross-beam assembly; dynamic characteristics; optimization design

Design and experimental Research of Automatic Tightening Method of Rubber Strip on the Side of Office Screen Panel

Ruonan Wang , Liang Han*,Jinghui Peng, Rui Peng

School of Mechanical Engineering Southeast University Nanjing211189, China
Corresponding author E-mail: melhan@seu.edu.cn

Abstract. In modern office space, screen desks are used worldwide for its user-friendly spatial segmentation. The screen panel is an important component of the screen desk. The tightening of rubber strips on the side of screen panel is currently manually tightened. In this paper, for the problem of manual tightening of the rubber strip on the side of the office screen panel at home and abroad, an efficient method for automatic tightening of rubber strips on the side of office screen panel is proposed .The process includes flexible fixing of the screen panel, unwinding of the strip, feeding of the strip, pressing of the strip and cutting of the strip. Based on this, the related mathematical model analysis, mechanism design is carried out. Then, this paper analyzes the multi-motor synchronous control task used in the automatic tightening method of the side rubber strip on the side of the office screen panel. Through the comparison of PID, neuron PID control algorithm, the simulation results of MATLAB show that the neuron PID control is robust , In order to verify the rationality of the method, an automatic tightening experimental device was built and the relevant tightening experiment was carried out. The preliminary experimental results show that the design of the automatic tightening method for the rubber strip on the side of the office screen panel is reasonable and feasible. The proposed method provides a design basis for practical production device of automatic tightening of the rubber strip on the side of the office screen panel.
Keywords: tightening of rubber strip; mechanism design; neuron PID; tightening experiment

Wednesday 4th Dec 2019

9.30 - 10.50 am Session 6 Sensing Methods and Actuation

Dynamic simulation of linear objects with friction on 2D surface

Benjamin Maier, Marius Stach, and Miriam Mehl
Institute for Parallel and Distributed Systems, University of Stuttgart, Germany

Abstract. Handling flexible, one-dimensional objects like tubes, cables, strings or ropes usually requires to quickly predict the dynamic behavior of the object. Tasks of robotic arms manipulating flexible one-dimensional objects often take place on a workspace with a flat surface. Here, the dynamic behavior of the bending, flexible object is governed by friction between the object and the surface. In this paper, we formulate a model based on differential geometry that predicts the dynamic behavior of an elastic one-dimensional object on a 2D workspace with friction. Orientation and position of one end of the object are prescribed by the gripper movement. We present algorithmic and numeric optimizations of the computation and provide details on our parallel implementation which achieves real-time performance. We showcase different features of the model and validate the modeling approach by comparison with experiments.

Keywords: Handling of deformable linear objects, dynamics, friction, real-time simulation

A highly adaptable embedded platform for sEMG acquisition

Zhao Zeming, Lv Bo, Sheng Xinjun*, and Zhu Xiangyang
State Key Laboratory of Mechanical System and Vibration, Shanghai Jiaotong University, 800 Dongchuan Road, Minhang District, Shanghai, China
xjsheng@sjtu.edu.cn
http://bbl.sjtu.edu.cn

Abstract. Generally, it is a contradictory between portability and high-performance in the field of high-density surface electromyography (HD-sEMG) acquisition equipment. Moreover, it limits the development of HD-sEMG to a great extent. In this paper, we have developed a portable device with high performance. The work focuses on the multi-level design of the system to balance the contradiction between the two to achieve a better result. A powerful but small HD-sEMG acquisition device and a versatile cross-platform PC software are developed to realize the system. Its acquisition device consists of a set of preamplifiers and a signal acquisition box. Preamplifiers guarantee the high quality of the signal during transmission. Signal acquisition box completes the analog-to-digital conversion of the signal with high precision and high sampling frequency and transmits it to the upper computer in a reliable way. Its upper software has an excellent user interaction interface, which can conveniently control various functions of the embedded device. Besides, it has various functions such as real-time display, data triggering, offline data saving, and so on. The feature of cross-platform allows all users to use it on any PC. At last, we test the entire system and use the standard algorithm to predict the finger movements to prove the system meets the specific requirements of sEMG collection.
Keywords: EMG acquisition device, HD-sEMG

Challenges in robotic soft tissue manipulation - Problem identification based on an interdisciplinary case study of a teleoperated painting robot in practise

M. Wnuk, F. Jaensch, D. A. Tomzik, Z. Chen, J. Terfurth, S. Kandasamy, J. Shahabi, A. Garrett, M. H. Mahmoudinezhad, A. Csiszar, W. L. Xu, O. Röhrle, A. Verl

Abstract.
Manipulating soft materials with robots is a complex and highly interdisciplinary research topic. Due to the involved underlying body of theories including material science, (bio-)mechanics, control, sensing as well as electrical and mechanical engineering, it is even difficult to provide a comprehensive problem formulation. In this paper, the problem of identifying challenges within the different research areas and between their respective interfaces is addressed with a practical case study. Based on a generalized sandbox scenario of a remotely controlled drawing robot, challenges with regard to the interdisciplinary research fields of robot control, architecture, sensor development, material modelling and interaction between rigid actuators acting on soft materials are investigated and analyzed. The major contribution of this paper is the explicit identification of unresolved research problems in the field and a quantification of the correlations between the involved disciplines. Furthermore, novel approaches on how to cir-

cumvent some of the found challenges in the future are proposed.

Keywords: soft material, problem formulation, manipulation, robot, teleopera-tion, case study

Modeling of lens based on dielectric elastomers coupling with hydrogel electrodes

Hui Zhang, Zhisheng Zhang*
School of Mechanical Engineering Southeast University Nanjing, China
Email: oldbc@seu.edu.cn, 103200003@seu.edu.cn

Abstract—At present, dielectric elastomer actuators (DEAs) have been utilized to tune liquid lenses. However, there are also some challenges, such as high actuating voltage and limited focus distance. This paper describes a lens system enabled by stretchable and transparent hydrogel electrodes; and voltages are applied on the layers of the hydrogel. The voltage-strain curve is calculated by using a model that accounts for the elastic constraint. When the thickness of a lens film is as a constant, the curvature of the soft–actuated lens becomes larger as the focal length increases. When the thickness increases, the focal length has a little change. Large actuated strain is attainable under the condition of large pre-stretch and thin hydrogel. A significant hysteresis is observed during cyclic loading and it becomes repeatable; the peak of the displacement is shifted from the peak of the voltage signal and occurs after it. To account for these viscoelastic phenomena, a constitutive model is developed by employing several dissipative non-equilibrium mechanisms. This modeling approach can be useful for control of viscoelastic soft-actuated lens.
Keywords—Transparent electrode; Stretchable hydrogel; Dielectric elastomer; Lens

11.15 - 12.15 Session 7 Agriculture Applications 2

Mechatronics for a LiDAR-based Mobile Robotic Platform for Pasture Biomass measurement

M. Sharifi1*, S. Sevier1, H. Zhang1, R. Wood1, B, Jessep1,
S. Gebbie1, K. Irie3, M. Hagedorn3, B. Barret2, K. Ghamkhar2
1Development Engineering, Lincoln Research Centre, AgResearch, Lincoln, New Zealand
2Forage Science, Grasslands Research Centre, AgResearch, Palmerston North, New Zealand
3 Red Fern Solutions Ltd, Christchurch, New Zealand
*mostafa.sharifi@agresearch.co.nz

Abstract. In-field measurement of pasture biomass and growth rate is imprecise and costly, limiting its research and commercial applications. In this paper, we present the mechatronic design and development process of a LiDAR-based mobile robotic platform for real-time non-invasive measurement of pasture vegetative biomass. This integrated mechatronic design and development process includes design specification, mechanical, electrical and control system design, as well as hardware and software development. Integrated application of CAD/CAM was applied to develop the 3D model of the robotic platform for rapid prototype development and manufacturing of the mechanical parts. The control system incorporates an embedded system to accommodate motion and operation control software, this is interfaced with the main processing computer, which runs the LiDAR measurement data processing algorithms. RTK-GNSS data is integrated with the LiDAR data to accurately map the scanned area. Our integrated approach provides a precise, non-destructive, and cost-effective system for real-time in-field measurement of pasture yield with highly anticipated scientific and commercial benefits.

Keywords: Biomass, Pasture, LiDAR, CAD, CAM, RTK-GNSS, Mechatronics design.

Vision guidance with a smart-phone

John Billingsley
University of Southern Queensland Toowoomba, QLD4350, Australia
john.billingsley@usq.edu.au

Abstract. A machine-vision-based guidance system is presented that can be hosted on a smartphone or tablet. This can provide economical vision guidance to an autonomous farm vehicle that is small. The original algorithm was developed at the turn of the millennium for a computing system of small capability. With advances in HTML5 it has now been updated and adapted. A link is provided to show the system in operation, from which the code can be downloaded for further research. As demonstrated, touch-screen operations have been provided to adjust the necessary parameters of camera alignment and row-spacing. has been cited many times, including six citations within the last two years. The simple strategy has now been encapsulated in the form of JavaScript code for a smart-phone or tablet and is demonstrated in action. A URL is given from which the source can be downloaded. With the addition of simple Bluetooth communication to a steering module, this means that precision automatic guidance along a row crop can now be applied to the smallest of mobile devices at a budget cost.
Keywords: Vision guidance, agriculture, tractor, JavaScript.

A high-speed camel dung collection machine

Samuel N. Cubero, Mohammad Badi, Mohamed Al Ali, Mohammed Alshehhi

Abstract. In the UAE (United Arab Emirates), camels are widely used in racing events and for 'desert safari' adventures. The dung from such animals can be used as a fertilizer for plants and gardens, and is currently being collected manually using shovels and wheelbarrows – a task that is considered by many workers to be tiring, tedious and laborious, especially in a desert region where temperatures typically range from 40 to 50 degrees C during the daytime in summer months. This paper describes the mechanical design, analysis and control of a mobile electric-powered machine that was designed to collect most kinds of litter and animal dung at very high speed. The current prototype that was designed and built at Khalifa University is self-propelled using a built-in drive motor. Thus, it does not need a large, expensive, powerful vehicle or farm tractor to power it, tow it or steer it, like most conventional litter collection machines. Also, the machine's steering direction and speed of movement can be manually controlled by an operator. Important design objectives and mechanical design details are explained, along with calculations, to show how the conveyor and drive motors were selected. Some of the serious technical problems encountered, and

possible solutions, are also discussed. With some minor additions, this kind of machine can be modified for future use as a fully autonomous mobile robot or a semi-autonomous remote-controlled machine that can be used for litter collection for major beaches and grass-covered parks. It could also be used for manually or automatically collecting most kinds of large animal dung, on horse and camel ranches, and on open fields where bovines and dairy cows dwell.
Keywords: dung, manure, litter, collecting, animal, camel, horse, mechatronic, machine, conveyor, rake

12:15 - 12.45 pm Introduction to M2VIP 2020

2:00-2:45 Session 8 Keynote 2

3:10- 4.10 pm Session 9 Industrial Processes and Products

A General Monitoring Method for the State of Spandex Based on Fuzzy Evaluation and Its Application

Limiao Gu1,1, Yan Wen1, Yu Zhang1, Weijie Chu1, Yunde Shi1, Fang Jia1*,
1 Department of Mechanical Engineering, Southeast University,
211189 Nanjing, China
Corresponding author E-mail: jfang@seu.edu.cn

Abstract. This paper aims at the problem of timely detection of spandex breakage during spinning. On this foundation, a general monitoring method for the state of spandex based on fuzzy evaluation is proposed. An on-line system which judges the state of spandex by means of the godet rollers is designed. Using the monitoring algorithm and process identification algorithm, the state of the godet rollers and the current production process are identified with the help of historical data. Meanwhile, an adaptive communication algorithm is presented to make the system suit the direction of data transmission. The experiment shows higher accuracy and effects of the given monitoring method compared to traditional methods, which enables real-time monitoring of spandex to ensure the quality of core spun yarn.
Keywords: spandex, core spun yarn, fuzzy evaluation, monitoring algorithm

Study on the Type Identification of Cheese Yarn Based on Low-Resolution Pictures

Xiaolong Liu1,1 , Ran Hu1, Yan Wen1, Yu Zhang1, Weijie Chu1, Zhisheng Zhang1, Fang Jia1*
1 Department of Mechanical Engineering, Southeast University,
211189 Nanjing, China
Corresponding author E-mail: jfang@seu.edu.cn

Abstract. This paper aims at the problems of identification of cheese yarn in low resolution pictures. On this basis, a cheese yarn label identification system based on machine vision technology is proposed. By using the improved Hough algorithm, the background of the label is removed. On the basis of transforming images from RGB color space to HSV color space, the color-histograms of the label are obtained. Then, the weights of the neural network are obtained by training the neural network with

histogram information, and the neural network is used to judge whether the types of yarns are the same as the samples. As verified by experiments, the stability and accuracy of the method can meet the practical application requirements and could effectively identify the types of cheese yarns.
Keywords: color feature extraction; color-histograms; cheese yarn; neural network

Research on high feeding speed system of L-valve rods based on two-in-one device

Shiwei Cheng, Liang Han*, Kai Yu, Rui Peng
School of Mechanical Engineering Southeast University Nanjing 211189, China
Corresponding author E-mail:melhan@seu.edu.cn

Abstract. In the production line, parts feeding plays an important role of it, which orientates the parts according to a specific attitude for the subsequent processing. L-valve rods (LVR) are key components of spray equipment, at present most enterprises adopt ordinary vibratory feeders for feeding, but the efficiency is too low to meet the production speed requirements of the enterprise. Besides, due to the small structure and irregular shape of the parts, the qualified rate of products is not high. In order to solve the above problems, in this paper a high feeding speed system of LVR based on a two-in-one device is presented, and the key parts of the machine are designed and analyzed, which includes vibratory feeders, two-in-one device, positioning device and cylinders, in which the two-in-one device is composed of left and right two curved fixed sections, two vertical moving sections and one vertical fixed section. The function of the two-in-one device is used to convert the two outputs of two vibratory feeders into one output. First, the two vibratory feeders sort the disorderly LVR and complete the directional feeding, then LVR sequentially enter the left and right curved fixed sections, vertical moving sections. When the sensor detects that LVR in vertical moving sections are full, the left moving section channel is shifted to the right pushed by the left cylinder, so that the outlet of the moving section channel is aligned with the inlet of the vertical fixed section channel. After LVR going through the vertical fixed section channel, the left cylinder returns, and the right cylinder performs correspondingly. Finally the prototype is built, based on which the feeding speed test is carried out. The experimental data shows that the mechanism can output the disordered LVR in the correct direction and accurately position it to the specified position. At the same time, the feeding efficiency can be greatly improved to meet the feeding speed requirement.
Keywords: L-valve rods, feeding machine, high speed, two-in-one device

Thursday 5th

9:30am – 10:30 Session 10 Robotics and Devices 2

A Scene Feature Based Eye-in-Hand Calibration Method for Industrial Robot

Guoshu Xu and Yonghua Yan
School of Mechanical Engineering Shanghai Jiao Tong University Shanghai, China
yhyan@sjtu.edu.cn

Abstract. The industrial robot eye-in-hand calibration has been a hot research topic for a long time. When the RGB-D camera is mounted on the end-effector of the robot, the pose of object can be obtained by the camera, then the pose transformation from camera coordinate system to robot coordinate system is realized by the hand-eye transformation matrix. Traditional marker-based calibration methods are so complex to operate and time-costing that are not suitable for actual application in industrial field. In this paper, with the aim of reducing the calibration complexity and time, a novel eye-in-hand calibration method based on scene feature is proposed. At First, the image sequence obtained by small motion of camera is processed by ORB feature extraction and Bundle Adjustment (BA), and the camera intrinsic parameters are solved. Then, ORB feature and Perspective-n-Point (PnP) is used to solve pose transformation between multi-view images, followed by BA optimization. Finally, the hand-eye transformation matrix is obtained by solving the corresponding relations between pose transformations of multi-view images and pose transformations of the robot. Since our method needs no man-made marker, it is thought as an auto calibration method and can be done many times during the operation process. Experiments show that compared with existing calibration methods, our method is simple, fast with high accuracy and more suitable for industrial application.
Keywords: industrial robot, RGB-D camera, Eye-in-hand calibration, bundle adjustment (BA)

A trajectory planning for a five degree of freedom vertically downward assistive feeding robotic arm using linear path with parabolic blend

Priyam A. Parikh*1, Dr. Keyur D. Joshi2 and Dr. Reena Trivedi3
1Research Scholar, NirmaUniversity,India,
2Assistant Professor, Ahmedabad University, India,
3Associate Professor, Nirma University, India
1,2{myrobo56,joshikeyurd}@gmail.com,
3reena.trivedi@nirmauni.ac.in

Abstract. In the recent era, sales of house hold robots are increasing due to their lesser cost and versatility. Trajectory planning plays a vital role in deciding the motion sequence of multiple degree of freedom robotic arms Trajectory planning can be done by using either using Cartesian scheme or using joint space scheme Joint space scheme mainly deals with angular displacement velocity and acceleration, which is important for a feeding robot. Parabolic blends can control infinite acceleration and deceleration at the starting and end points of the trajectory, but it not able to control the instantaneous velocity and acceleration in the middle of the trajectory. Apart from that, it is not able to provide zero acceleration at the beginning and at the end point. In order to overcome this problem, this paper proposes a novel method of designing a trajectory using full cycloid and semi function. This paper also compares trajectories generated by linear trajectory with parabolic blends, fully cycloid function and semi cycloid functions along with angular velocity and acceleration. A vertically downward five degree of freedom T-R-R-R-R type robotic arm is considered in this work. The camera attached with the arm base, captures an image of a seated person, locates the destination point, and sends it to the microcontroller. The objective of the robotic arm is to feed the physically challenged people. By using the proposed method of trajectory planning, a smoother trajectory compared to LSPB (linear segment with parabolic blend) along with finite acceleration can be achieved.

Structure design and closed-loop control of a modular soft-rigid pneumatic lower limb exoskeleton

Jiangbei Wang1,2, Yanqiong Fei*,1,2

Abstract. This paper presents the structure design and closed-loop control of a novel modular soft-rigid pneumatic exoskeleton for lower limb. The proposed exoskeleton consists of three soft-rigid modules and one foot module. Each soft-rigid module consists of one active soft hinge and one rigid link. Two different modules can be joined by the complementary convex and concave surfaces at their two ends. Each soft hinge is made of and actuated by a customized bidirectional Curl Pneumatic Artificial Muscle (CPAM). Each of the modules integrate two pressure sensors, two flex sensors and an Inertia Measurement Unit (IMU) for estimating the bending angles of the soft hinges. An external electro-pneumatic control system is developed. The closed-loop control system consisting of pressure servos and position controllers is designed to control the bending angles of the exoskeleton hinges. Experiment shows that the proposed soft-rigid exoskeleton has good motion controllability.
Keywords: soft robot, module, soft-rigid exoskeleton, closed-loop control

10:55 - 11:55 pm Session 11 Robotics and Devices 3

Discussion of soft tissue manipulation for the harvesting of Ovine offal

Qi Zhang1, Weiliang Xu1*, Zhisheng Zhang2, Martin Stommel3, Alexander Verl4
1 Department of Mechanical Engineering, The University of Auckland, Auckland, New Zealand
qzha862@aucklanduni.ac.nz, p.xu@auckland.ac.nz
2 School of Mechanical Engineering, Southeast University, Nanjing, China
oldbc@seu.edu.cn
3 Department of Electrical and Electronic Engineering, Auckland University of Technology, Auckland, New Zealand
mstommel@aut.ac.nz
4 Institute of Control Engineering (ISW), Stuttgart University, Stuttgart, Germany
alexander.verl@isw.uni-stuttgart.de

Abstract. Ovine offal is a significant export co-product of meat processing in New Zealand. There is currently no automatic system available for its sorting. In this paper, we first analyze the requirements for robotic sorting of internal sheep organs, including the particular needs of the food industry, the components required for the robotic sorting system and the manipulation tasks. Current practices in the animal slaughtering and processing industry are followed. And then a robotic system for ovine offal sorting is proposed, and the problems to be solved are summarized. Next, we discuss the current methods and theories to solve the above problems and give potential solutions. Finally, the conclusion and future work of ovine offal harvesting are given.

Keywords: soft tissue manipulation, ovine offal sorting, robotic manipulation, modelling

Robots for Complex Cutting Operations in Beef Processing

F.Border, K.Khodabandehloo and P.Brett

Abstract The processing of meat is complex for automation machinery as there is a wide range in variation of presentation of deformable meat workpieces. Markets at the other end of the chain require outcomes with less variation and other qualities on the final presentation. Tasks are in cutting and handling. In a few automation examples in poultry and pork processing there has been measured success where constraints can be applied to minimise the sensitivity of initial variation on the final outcome. This paper describes research with the aim to produce robot solutions able to cope with the variations expected in the wider cutting operations in processing beef. The trimming and slicing of high value beef primals presents further complications that if solved will enable new tier of automation techniques and the shift in supporting enabling technology needed for solutions in a wide range of cutting operations. The focus of the paper in cutting to tissue interfaces illustrates the challenges in sensing and decision functions to control a cutter on the correct trajectory within the meat workpiece and reports on merits of some possible approaches in practice.
Keywords Robotics, meat, cutting operations, sensing

Microbial Electrochemical Sensors to Detect Volatile Fatty Acids in Anaerobic Digestion

Andrew Hill

Abstract Testing of volatile fatty acids (VFAs) is important to control the operation of anaerobic digesters that produce renewable biogas energy, yet VFAs are usually measured manually, with few automated VFA monitoring systems. Current manual methods are time consuming and often costly [1, 2]. Microbial electrochemical sensors (MESe) are emerging as a simpler low-cost alternative to detect VFAs, provided that some key challenges can be overcome such as VFA specificity, system fouling, cross sensitivity, biofilm maturity, signal degradation, and response times [3, 4]. Mechatronics is central to the control of systems, and machine perception of the state of the process is key. This paper provides an overview of online detection of VFAs using a MESe, and will focus on improving accuracy and response time, and minimising fouling of the sensor. At the conference the process will be described through practical examples.
Keywords Online Microbial Electrochemical Technologies · Acetate · High Strength Wastewater

2:00 - 3:15 pm Visit to the Centre for Agricultural Engineering

4:00pm End of Conference

AUTHOR INDEX

NOTES:

Mechatronics and Machine Vision in Practice
FOES, University of South Queensland
Toowoomba QLD 4350 Australia

ISBN 978-1-7138-0209-9

9 781713 802099

CURRAN ASSOCIATES INC.
proceedings.com